Cambridge Primary
Science
Second Edition
Workbook 1

Series editors:
Judith Amery
Rosemary Feasey

Boost

HODDER
EDUCATION
AN HACHETTE UK COMPANY

Cambridge International copyright material in this publication is reproduced under licence and remains the intellectual property of Cambridge Assessment International Education.

Registered Cambridge International Schools benefit from high-quality programmes, assessments and a wide range of support so that teachers can effectively deliver Cambridge Primary. Visit www.cambridgeinternational.org/primary to find out more.

Acknowledgements
The Publishers would like to thank the following for permission to reproduce copyright material.

Photo acknowledgements
p. 8 *tl, cr,* **p. 31** *tl, cr,* **p. 43** *tl, cr,* **p. 50** *tl, cr,* **p. 58** *tl, cr,* **p.62, p. 65** *tl, cr,* **p. 73** *tl, cr,* **p. 80** *tl, cr* © Stocker Team/Adobe Stock; **p. 60** *bl* © Jenesesimre/Adobe Stock; **p. 62** *br* © Tomislav Forgo/Fotolia; **p. 74** *tl* © Alex Okokok/Adobe Stock; **p. 74** *tc* © Robert/Adobe Stock; **p. 74** *tr* © Dottedyeti/Adobe Stock.

t = top, *b* = bottom, *l* = left, *r* = right, *c* = centre

Every effort has been made to trace all copyright holders, but if any have been inadvertently overlooked, the Publishers will be pleased to make the necessary arrangements at the first opportunity.

Hachette UK's policy is to use papers that are natural, renewable and recyclable products and made from wood grown in well-managed forests and other controlled sources. The logging and manufacturing processes are expected to conform to the environmental regulations of the country of origin.

Orders: please contact Hachette UK Distribution, Hely Hutchinson Centre, Milton Road, Didcot, Oxfordshire, OX11 7HH. Telephone: +44 (0)1235 827827. Email education@hachette.co.uk Lines are open from 9 a.m. to 5 p.m., Monday to Friday. You can also order through our website: www.hoddereducation.com

© Rosemary Feasey, Deborah Herridge, Helen Lewis, Tara Lievesley, Andrea Mapplebeck, Hellen Ward 2021

First published in 2017

This edition published in 2021 by

Hodder Education,

An Hachette UK Company

Carmelite House

50 Victoria Embankment

London EC4Y 0DZ

www.hoddereducation.com

Impression number   10 9 8 7 6 5 4

Year   2025 2024 2023 2022 2021

Cover illustration by Lisa Hunt, The Bright Agency

Illustrations by Alex van Houwelingen, Ammie Miske, James Hearne, Jeanne du Plessis, Natalie and Tamsin Hinrichsen

Typeset in FS Albert 17/19 by IO Publishing CC

Printed in the United Kingdom

A catalogue record for this title is available from the British Library.

ISBN: 9781398301450

MIX
Paper from
responsible sources
FSC™ C104740

Contents

Biology

Unit 1 Alive or not alive?   4
Self-check   8
Unit 2 Plants   9
Self-check   18
Unit 3 Ourselves   19
Self-check   31

Chemistry

Unit 4 Material properties   32
Self-check   43

Physics

Unit 5 Forces   44
Self-check   50
Unit 6 Sound   51
Self-check   58
Unit 7 Electricity and magnets   59
Self-check   65

Earth and space

Unit 8 Planet Earth   66
Self-check   73
Unit 9 Earth in space   74
Self-check   80

Unit 1   Alive or not alive?

Is it alive or not alive?

1   Draw a circle around the things that are alive.
One has been done for you.

giraffe   lizard   human   car   chair   mug   tree   plant

2   Draw something that is **alive** and something that is **not alive**.

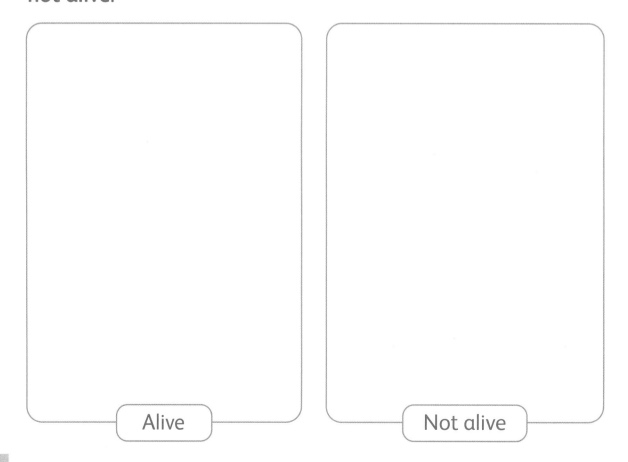

Alive   Not alive

A living thing

1 Camila has a pet bird called Bluey.

What does Bluey do that tells you he is alive?

_____

_____

_____

_____

2 Complete the sentences. Use these words:

(grown) (moves) (eats) (breathes)

a Bluey _____ around his cage.

b Bluey _____ when he is hungry.

c Bluey _____ through two holes on his beak.

d Camila started looking after Bluey when he was a baby.

   Bluey has now _____ bigger.

Find out!

1 Find out if a dandelion plant is alive.
Tick (✓) **yes** or cross (✗) **no**.

Question	Yes ✓	No ✗
a Does the dandelion plant grow?		
b Does it make new dandelion plants?		
c Does it need food?		

dandelion plant

2 Is a dandelion plant alive? Tick (✓) **yes** or cross (✗) **no**.

Yes ☐   No ☐

3 Find out if a kettle is alive.
Tick (✓) **yes** or cross (✗) **no**.

Question	Yes ✓	No ✗
a Does the kettle move?		
b Does it make new kettles?		
c Does it need food?		

kettle

4 Is a kettle alive? Tick (✓) **yes** or cross (✗) **no**.

Yes ☐   No ☐

Find living things

1 Help Joe find the living things in his street.
Draw circles around the living things.

Unit 1 Alive or not alive?

Self-check

 I can do this.

 I can do this, but I need to keep trying.

 I can't do this yet.

See how much you know!

What can I do?	😉	😐	🙁
1 I can sort things into groups.			
2 I can say what is a living thing.			
3 I can say how I know something is a living thing.			
4 I can say what is alive or not alive.			
5 I can say why humans are living things.			
6 I can say why plants are alive.			
7 I can say why animals are alive.			
8 I can say what an animal needs to stay alive.			
9 I can sort things into **alive** and **not alive**.			

I need more help with:

_____

_____

_____

Plant parts

1 Label the parts of the plant. Use these words:

leaf	stem	flower	root

2 Look at each word. Cover each word. Write each word. Check your spelling. Write the word again.

You have three chances.

a roots _____   _____   _____

b stem _____   _____   _____

c leaf _____   _____   _____

d flower _____   _____   _____

Parts of a plant

1 Maris made a plant wristband. She stuck parts of a plant on it.
Draw lines to label the parts of the plant.

flower

petal

leaf

roots

stem

2 Design your own plant.
Label the parts of your plant.
Name your plant.

flower   leaf

stem   roots

My plant is called _____.

A plant hunt

1 Bashir looked for plants in his school grounds.

a Go on a plant hunt in your school grounds.

b Draw three plants that you find.

c Name the plant and label the parts.

Plant name:	Plant name:	Plant name:

Make a model plant

1 Make a model of a plant.
Use recycled materials.

  a Draw what your model plant will look like.

  b What materials will you use for each part
  of the plant?

  c Label the parts of the plant.

2 Show your model to a partner. Ask them to say:

  • something good about your model

  • how you can make your model better.

Grow seeds

Gemi planted a seed. She kept a diary of how the seed was growing.

Below each picture, write down which part of the plant has grown.

Day 1

_____

_____

_____

Day 4

_____

_____

_____

Day 7

_____

_____

_____

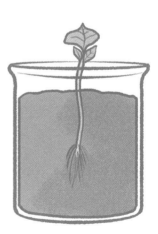

Day 8

_____

_____

_____

What do seeds need to grow?

1   Help Zara find out if seeds need water to grow.

Do seeds need water to grow?

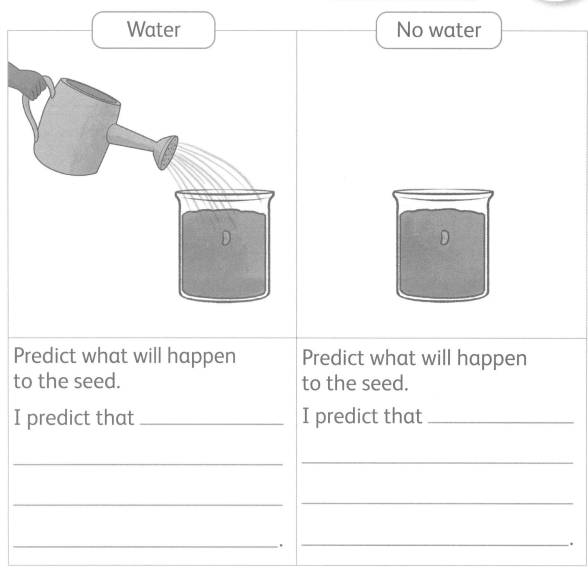

Water	No water
Predict what will happen to the seed.	Predict what will happen to the seed.
I predict that _____	I predict that _____
_____	_____
_____	_____
_____ .	_____ .

2   Predict what will happen to the seeds after three weeks.

I predict that _____

_____

_____

_____ .

What do plants need to stay healthy?

1 Gemi's teacher gave her two plants.

Gemi put one plant in a dark cupboard.

She put the other plant on the windowsill in the light.

Gemi kept a diary.

Do plants need light to grow?

	Day 1	Day 3	Day 5	Day 7	Day 9
Plant in the light	no change	green leaves	green leaves plant growing taller	green leaves thick stem	green leaves thick stem looks healthy
Plant in the dark – no light	no change	leaves getting yellow	yellow leaves very long stem	pale, yellow leaves thin and long stem	very pale, yellow leaves thin, long and floppy stem does not look healthy

Complete the sentences.

a On day 5 the plant in the dark had _____ leaves.

b On day 7 the plant in the light had a _____ stem.

c On day 7 the plant in the dark had a _____ stem.

d On day 9 the plant in the _____ was healthy.

How to measure bean plants

1 Anton and Bashir grew a bean plant.
They measured the plant every two days.

Here are their results.

| day 1 | day 3 | day 5 | day 7 | day 9 |

Answer these questions.

a How many cubes tall was the plant on day 3? _____

b How many cubes tall was the plant on day 5? _____

c On which day was the plant the tallest? _____

d On which day was the plant three cubes tall? _____

Plant words

1 Learn to spell these plant words. Ask someone to test you.

a (plant)   b (stem)

c (leaf)   d (flower)

e (tree)   f (roots)

2 Dani has mixed up six plant words. Sort each word so that it is correct. Write the correct word. Use the words in question 1 to help you.

a (mste)   b (lanpt)

_____   _____

c (lfea)   d (eert)

_____   _____

e (toors)   f (rewolf)

_____   _____

Unit 2 Plants

Self-check

 I can do this.

 I can do this, but I need to keep trying.

 I can't do this yet.

See how much you know!

What can I do?			
1 I can make a prediction about what I think will happen.			
2 I can use a magnifying glass.			
3 I can record my observations in a table.			
4 I can say what happened.			
5 I can say why plants are living things.			
6 I can name the different parts of a plant.			
7 I can say what seeds need to grow.			
8 I can tell someone how to grow seeds into plants.			
9 I know plants need water and light to grow.			
10 I can name parts of plants in the school grounds.			

I need more help with:

_____

_____

_____

Parts of the body

1 Label the picture. Draw a line from each word to the correct part of the body.

head

shoulder

hand

leg

ankle

elbow

ear

wrist

arm

knee

foot

2 **a** Find out where each of these parts are on the body.

neck   chin   fingers

b Write labels for these parts on the picture in question 1.

Who has the biggest hand?

Jack, David and Zara wanted to find out who had the biggest hand.

This is what they did:

1 Count the marbles in each hand.

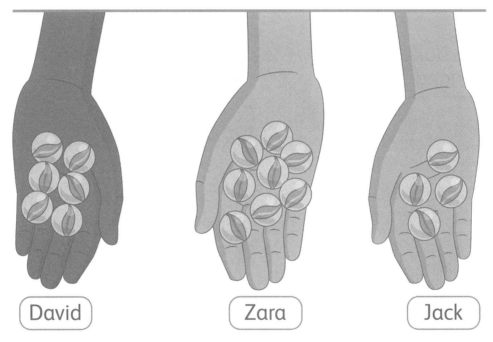

David   Zara   Jack

a Who has the biggest hand? _____

b Who has the smallest hand? _____

c How many marbles did David pick up? _____

2 Work in a group. Do what the learners did to find out who has the biggest hand in your group.

What size are your shoes?

The children wanted to find out what size shoe learners wear.

They made this block chart.

Number of children				
6				
5				
4		■		
3		■		
2	■	■		
1	■	■	■	
	9	10	11	12
	Shoe size			

1 Look at the chart and answer these questions.

a How many children wear size 9 shoes? _____

b How many children wear size 10 shoes? _____

c How many children wear size 12 shoes? _____

d What size shoe do you wear?

I wear size _____ shoes.

Shoe sizes in different countries												
UK	9	9½	10	10½	11	11½	12	12½	13	13½	1	1½	2
US	9½	10	10½	11	11½	12	12½	13	13½	1	1½	2	2½
EU	26	27	27½	28	28½	29	30	30½	31	31½	32½	33	34
cm	16	16½	17	17½	17½	18	18½	19	19	19½	20	20½	21

2 Find out who else in your class has the same size shoe as you.

The optometrist

1 Mario goes to see the optometrist. What does an optometrist look after? Circle the correct word below.

| teeth | hair | eyes |

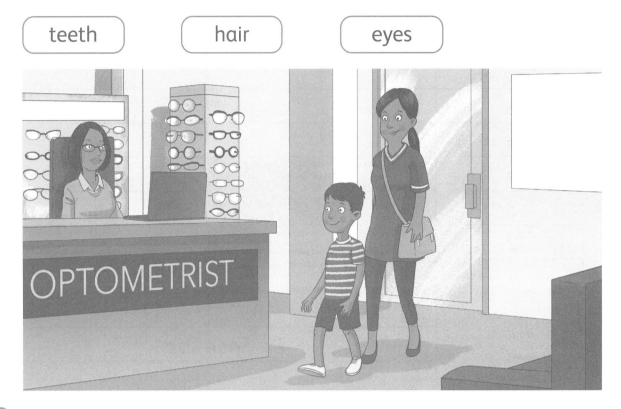

2 You can get new glasses at the optometrist.

Design a pair of glasses that you would like to wear.

My glasses design

Where are our senses?

1 **a** Draw a picture of yourself.

b Label your five senses. Use these words:

smell   touch   sight   hearing   taste

2 Match the sentence to the correct word.
One has been done for you.

a You use this for your sense of smell.

b You use this for your sense of touch.

c You use these for your sense of sight.

d You use these for your sense of hearing.

e You use this for your sense of taste.

ears

eyes

tongue

skin

nose

23

Our senses

1 Write a sentence about what you can do with each sense.
One has been done for you.

a smell   I can smell flowers with my nose.

b sight   _____

c touch   _____

d hearing _____

e taste   _____

2 Label the parts of the eye. Use these words:

eyebrow   eyelash   eyelid

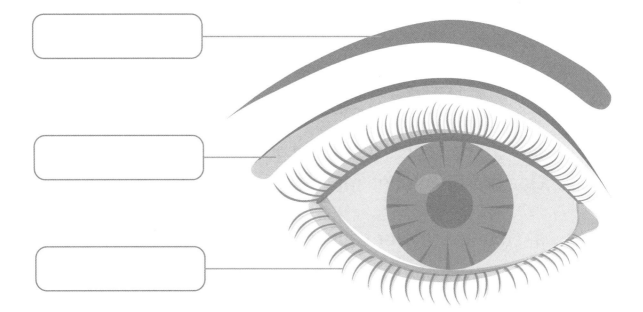

Find the differences!

1 Sight is one of the five senses. Write the names of the other four senses.

_____   _____

_____   _____

2 a Find six differences between the pictures.

   b Draw a circle around all the differences you can see in picture 2.

picture 1

picture 2

How big are your hands?

1

a   Is your hand bigger or smaller than your partner's hand?

b   How can you use these squares to find out? What will you do?

c   Try out your idea.

<table>
<tr><td></td><td></td><td></td><td></td><td></td><td></td><td></td></tr>
<tr><td></td><td></td><td></td><td></td><td></td><td></td><td></td></tr>
<tr><td></td><td></td><td></td><td></td><td></td><td></td><td></td></tr>
<tr><td></td><td></td><td></td><td></td><td></td><td></td><td></td></tr>
<tr><td></td><td></td><td></td><td></td><td></td><td></td><td></td></tr>
<tr><td></td><td></td><td></td><td></td><td></td><td></td><td></td></tr>
</table>

2   What did you find out? Is your hand **bigger** or **smaller**?

Complete these sentences.

a   My hand is _____ than my partner's hand.

b   I know this because my hand is _____ squares

big. My partner's hand is _____ squares big.

Different smells

 a Which smells do you like?
Which smells do you not like?

b Draw things you like to smell
and things you do not like
to smell.

Things I like to smell

Things I do not like to smell

Nice smells

1 This pictogram shows smells that some learners like.

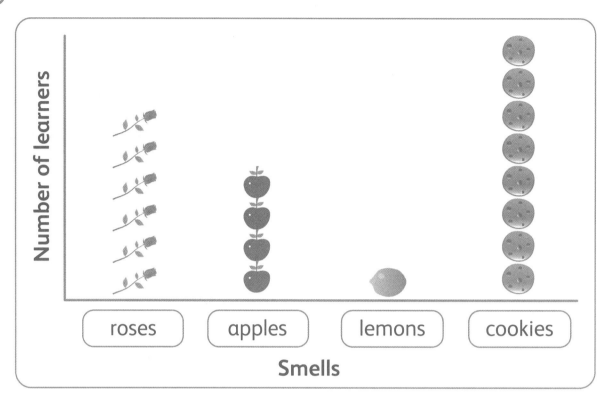

a How many learners like 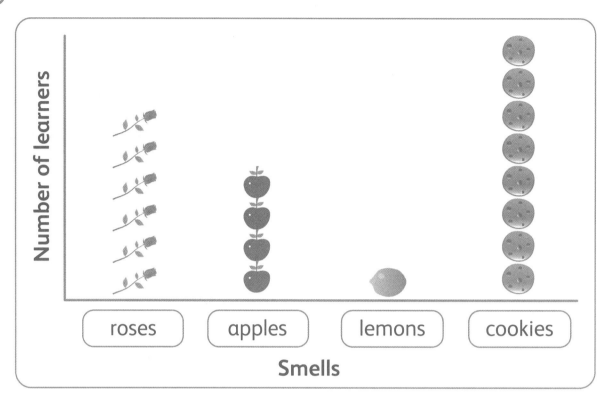? _____

b Which smell did one learner like? _____

c How many learners like ? _____

d Did more learners like or ? _____

2 Draw things that you can smell.

Indoor smells	Outdoor smells

Questions about taste

1 Annay asked his partner questions about taste.
Write three of your own questions about taste.

What is your favourite taste?

What taste do you like that is sour?

Do you like the taste of an apple or a pear?

a Question 1

_____

_____

b Question 2

_____

_____

c Question 3

_____

_____

2 Ask your partner your questions. Write down their answers.

a Question 1: _____

b Question 2: _____

c Question 3: _____

More about senses

1 Complete the sentences. Use these words:

five	salty
nose	touch
taste buds	see
smell	sweet

a   There are _____ senses.

b   The senses are sight, hearing, taste, _____

   and _____.

c   I use my eyes to _____.

d   If you hold your _____ when you eat
   an apple, you cannot taste it very well.

e   There are _____ on the tongue.

f   Taste buds help you to taste _____ , sour

   and _____ foods.

2 Complete the sentences. Choose from these words:

smell	sight	taste

hearing	touch

a   I think that the most important sense is _____.

b   It is important because _____

   _____.

Self-check

 I can do this.

 I can do this, but I need to keep trying.

 I can't do this yet.

See how much you know!

What can I do?			
1 I can measure and put my results in a table.			
2 I can say what is similar and different about me and someone else in my class.			
3 I can name parts of my body.			
4 I can name the five senses.			
5 I can name the part of the body for each sense.			
6 I can say that an optometrist is someone who uses science to make sure eyes are healthy.			

I need more help with:

_____

_____

_____

Materials all around us

1 Complete the sentences. Use these words:

(materials)   (objects)

a Everything around you is made from one or more

_____.

b All _____ are made from one or more materials.

2 a Find six different objects around you. Draw each object in one of the boxes below.

b What material or materials is each object made from?

Write the name of the materials under each picture.

_____	_____	_____
_____	_____	_____

Different textures

1 The labels on the wheel show different textures.
Draw objects that are made from materials with each texture.
One has been done for you.

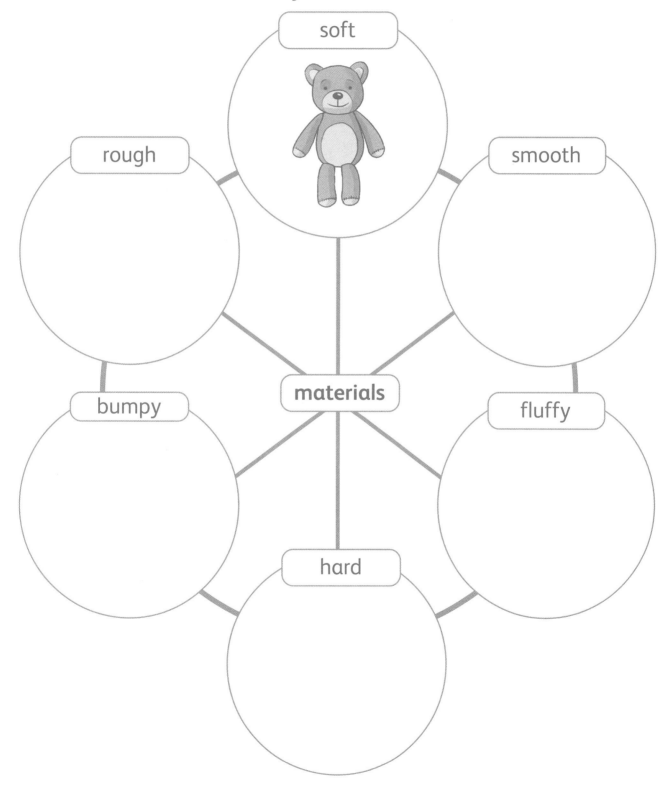

Materials in the kitchen

1 Draw an arrow pointing to each object in the picture. Write what material the object is made from. One has been done for you.

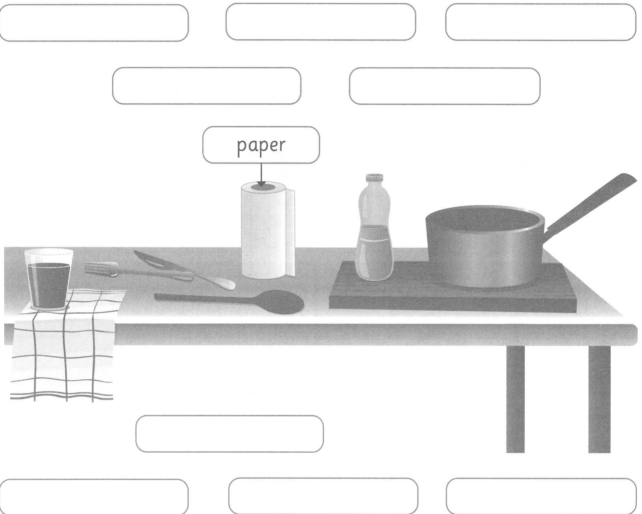

paper

2 Name one object in your kitchen that is made from these materials:

a metal _____

b glass _____

c plastic _____

d fabric _____

Metal objects

1 Cora collected these objects.

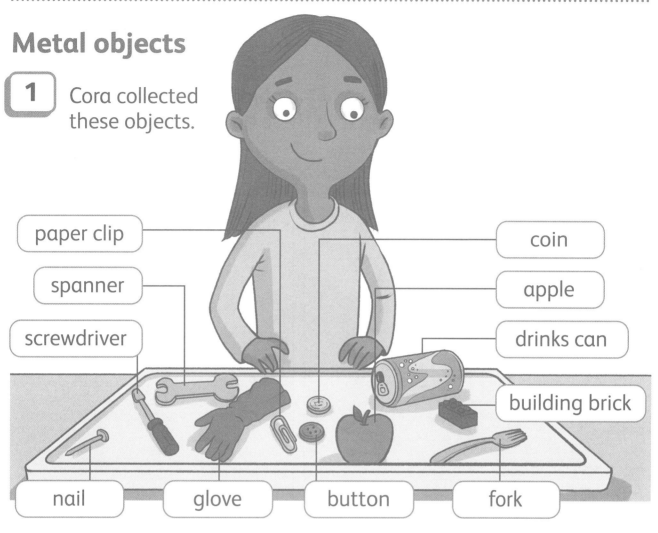

paper clip

spanner

screwdriver

coin

apple

drinks can

building brick

nail

glove

button

fork

Draw the objects that are **not** made from metal.

2 Tick (✓) the words that say what most metal materials are like:

furry · hard · fluffy · shiny

cold to touch · hot to touch

Plastic and wood

1 Look around your classroom. Find objects made from plastic and wood. Draw the objects in these boxes:

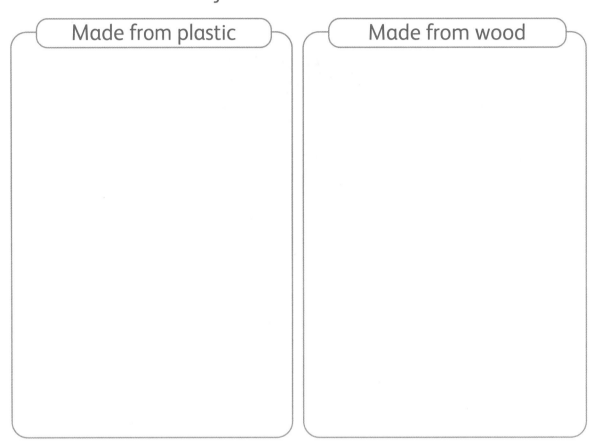

Made from plastic	Made from wood

2 What objects in your home are made from plastic?

Write three objects here:

a _____

b _____

c _____

Hint: Think about the different rooms in your home.

Glass objects

1 Read each sentence.
Tick (✓) true or cross (✗) false.

Sentence	True	False
a Glass can break.		
b Glass is soft.		
c Glass can be transparent (see-through).		
d All glass is red.		

2 Draw four objects in your home that are made from glass.

Made from glass

Clay

A potter is a person who uses clay to make things.

Potters know about materials. They use science to help them change the shape of the clay.

This potter is making a clay pot.

1 Tick (✓) the words below that say what clay feels like.

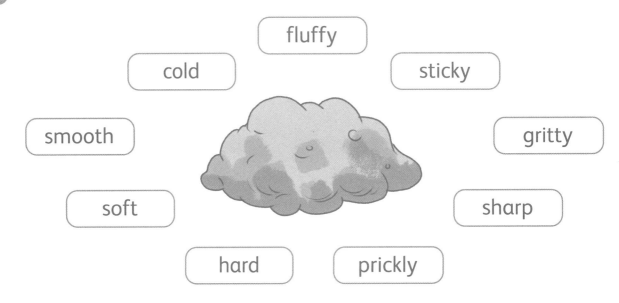

fluffy

cold

sticky

smooth

gritty

soft

sharp

hard

prickly

Which material is waterproof?

1 Which hats do you think are waterproof? Which hats might **not** be waterproof? Tick (✓) your predictions.

Material	I predict the hat is waterproof.	I predict the hat is not waterproof.
paper hat		
plastic hat		
fabric hat		
metal foil hat		

2 Test your predictions to find out if you are correct.

Waterproof test

David tested different materials
to find out if they were waterproof.

Here is what happened.

I wonder which
material is
waterproof?

Material	I predict it is waterproof.	What happened ...
plastic	yes	It did not let water through.
cotton fabric	no	It let water through.
paper	yes	It let water through.
tissue paper	yes	It let water through.

1 Tick (✓) the correct sentences.

a   The plastic was waterproof. ☐

b   The cotton fabric was waterproof. ☐

c   The paper is was waterproof. ☐

d   The tissue paper was waterproof. ☐

e   All David's predictions were correct. ☐

Stretch, squash, bend, twist

1 Cora is doing her exercises.

Write the correct word under each picture. Use these words:

| squash | twist | bend | stretch |

a

b

c

d

Change materials

1 Tick (✓) the objects below that twist, stretch, squash or bend.
One has been done for you.

Object	Twist	Stretch	Squash	Bend
balloon	✓	✓	✓	✓
metal fork				
tomato				
paper straw				
sponge				
brick				
rubber band				

Unit 4 Material properties

Self-check

See how much you know!

 I can do this.

 I can do this, but I need to keep trying.

 I can't do this yet.

What can I do?	😉	😐	🙁
1 I can say what happened during a test and if the predictions were correct.			
2 I can sort materials into groups.			
3 I can name different materials.			
4 I can say that objects are made from different materials.			
5 I can describe how different materials have different textures.			
6 I can describe how different materials have different properties.			
7 I can describe what different materials are like.			
8 I can tell someone how to change the shape of materials.			
9 I can describe how people use science when they use materials.			

I need more help with:

_____

_____

_____

Unit 5 Forces

How things move

1 Things move in different ways. Match the pictures to the correct words. One has been done for you.

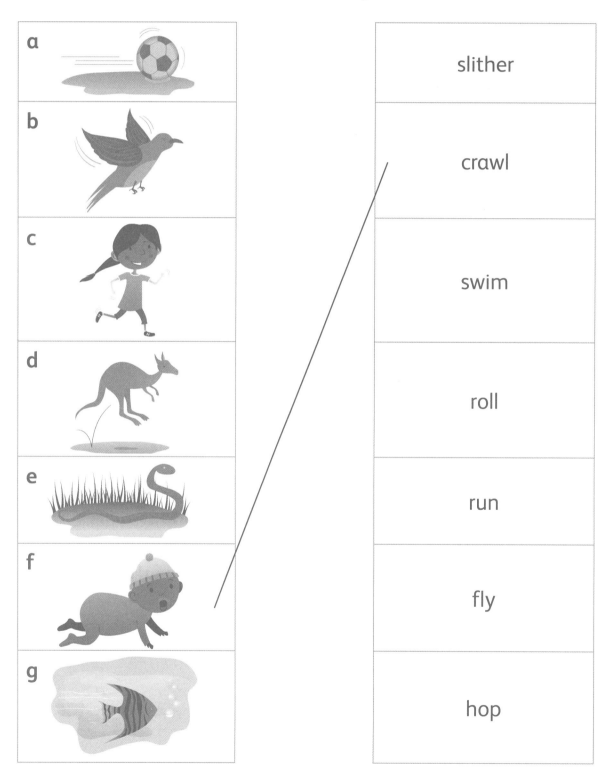

44

Twist, stretch, squash or bend

1 What is happening in these pictures? Use these words:

bending	squashing	stretching	twisting

a This boy is _____ his legs.

b These hands are _____ the dough.

c These hands are _____ the scarf.

d These hands are _____ the ruler.

2 Draw other objects that you can twist, stretch, squash or bend.

Pushes and pulls

1 Complete the sentences. Use these words: (pull) (push)

 a When you _____, you move the object away from you.

 b When you _____, you move the object towards you.

2 How do you move the objects below? Tick (✓) **push**, **pull**, or **both**. One has been done for you.

Object	Push to move	Pull to move	Both push and pull
doorbell	✓		
door			
swing			
shopping trolley			
toy			

Float and sink

1 Do you think the objects in the table will float or sink?

Tick (✓) **float** or **sink**. One has been done for you.

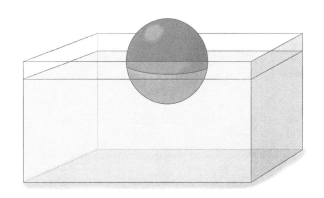

Object		Float	Sink
balloon		✓	
cricket ball			
stone			
glass marble			
metal nail			
plastic pot			
metal spoon			

Float or sink objects

1 Draw these objects in the water tank to show whether they float or sink.

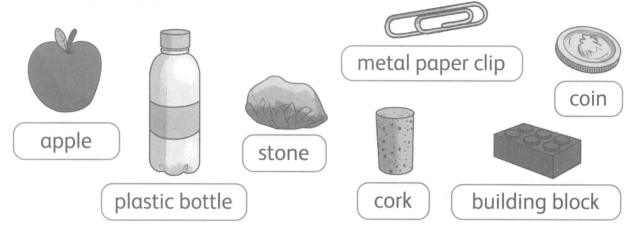

apple

plastic bottle

stone

metal paper clip

cork

coin

building block

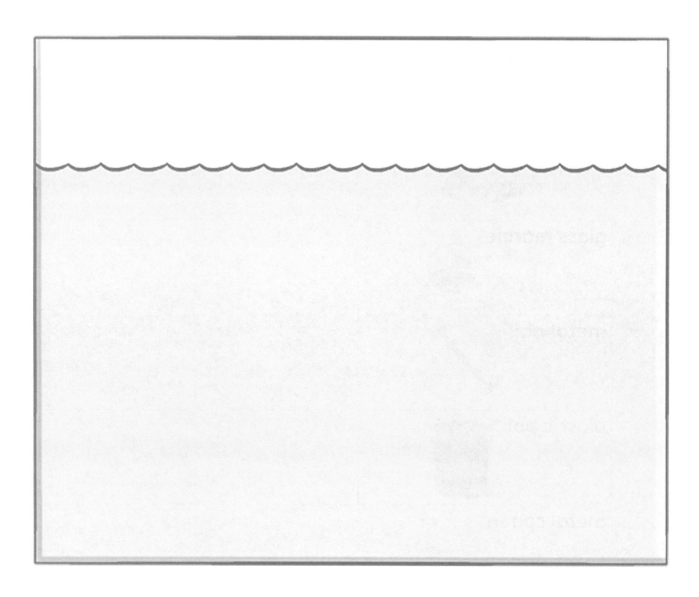

Design a sailboat

1 Design and make a sailboat.

a Draw your sailboat here:

My sailboat design

b Write labels for the materials you will use to make your sailboat.

c Will you use a push or a pull to make the sailboat move?

Unit 5 Forces

Self-check

 I can do this.

 I can do this, but I need to keep trying.

 I can't do this yet.

See how much you know!

What can I do?			
1 I can sort objects into groups.			
2 I can make predictions about what will happen.			
3 I can record and use simple tables.			
4 I can talk about how objects move.			
5 I can say what a push is.			
6 I can say what a pull is.			
7 I can say which objects float and which objects sink.			

I need more help with:

_____

_____

_____

Sounds I do not like

1 **a** Draw four sounds you do **not** like.

b Write a sentence about why you do not like each sound.

I do not like the sound because _____

_____

_____ .

I do not like the sound because _____

_____

_____ .

I do not like the sound because _____

_____

_____ .

I do not like the sound because _____

_____

_____ .

Sources of sound

1 Match the pictures to the correct sound. One has been done for you.

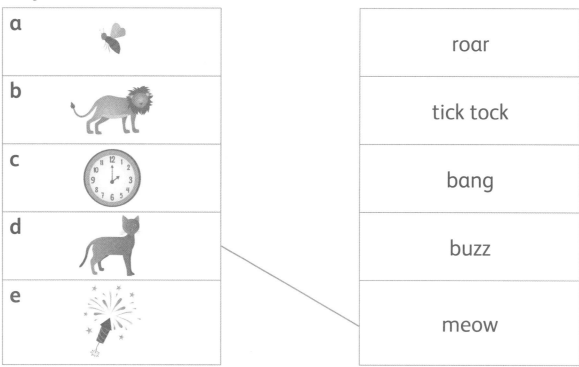

a	roar
b	tick tock
c	bang
d	buzz
e	meow

2 Draw a source of sound in each box. Write the name of the sound. One has been done for you.

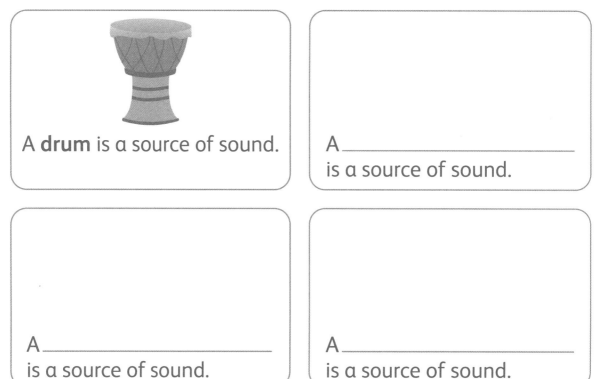

A **drum** is a source of sound.

A _____ is a source of sound.

A _____ is a source of sound.

A _____ is a source of sound.

Make sounds

1 Match the instrument with what you must do to make a sound. One has been done for you.

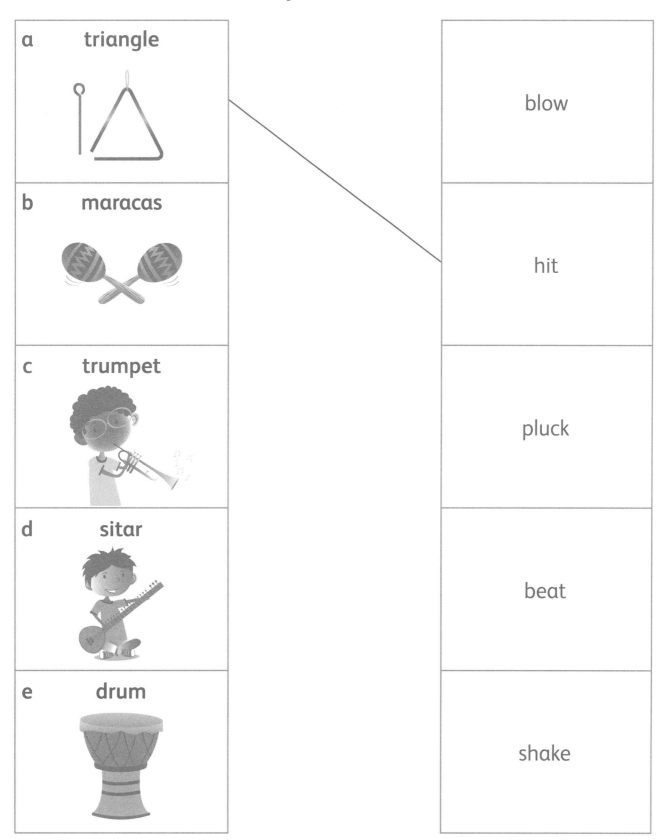

Amazing sounds

1 Imagine that you have an amazing bag of sounds.

  a Which sounds would you put into your bag?

  b Would they be sounds you like or sounds you do not like?

  c Draw and write the names of the sounds that you would put into your bag.

Make sound boxes

1 Taman and Malik made sound boxes. They put different objects into their boxes. They shook the boxes to make different sounds.

Here are objects they used:

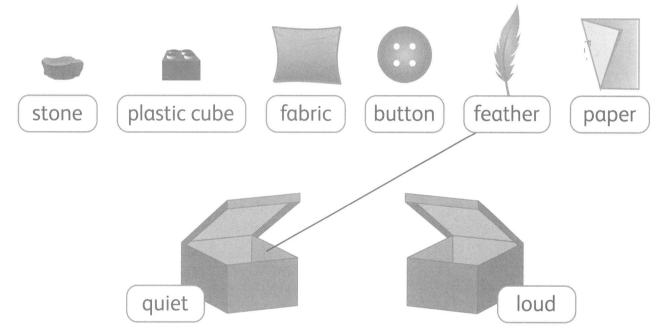

stone   plastic cube   fabric   button   feather   paper

quiet   loud

Draw a line from each object to the correct sound box. One has been done for you.

2 Draw four other objects that could go into a sound box. Think of objects that will make interesting sounds.

Sounds near and far

Layla made some sounds. Li listened to find out which object she could still hear from 20 steps away.

Here is what they found out:

Object	Could Li hear it?
1 feather shaker	no
2 tambourine	yes
3 cymbals	yes
4 whistle	yes
5 tapping sticks	no

1 Write the numbers of the objects.

   a Which sources of sound could Li hear? _____

   b Which sources of sound could Li not hear? _____

2 How can Layla make a louder sound with the tapping sticks?

_____

_____

Sounds get fainter

A group of learners wanted to find out how far away they could still hear sounds. One learner hit a drum while the others walked away. They stopped when they could not hear the drum.

This bar chart shows what happened. Use it to answer the questions below.

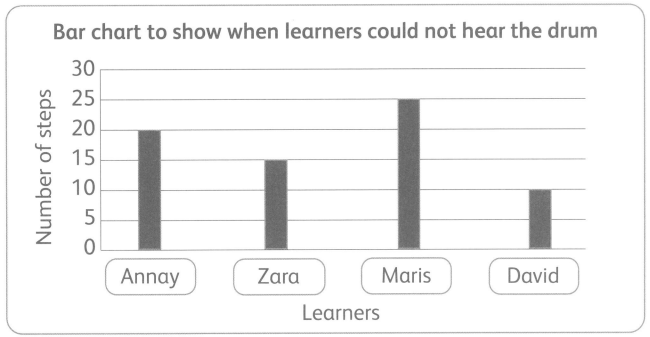

Bar chart to show when learners could not hear the drum

Number of steps (y-axis: 0, 5, 10, 15, 20, 25, 30)

Annay — 20
Zara — 15
Maris — 25
David — 10

Learners

1 a Who could hear the drum the furthest away?

_____

  b How many steps did Maris take until she could not hear

   the drum? _____

  c Draw a picture of the sound test the learners did. Use another piece of paper if you need to.

Unit 6 Sound

Self-check

 I can do this.

 I can do this, but I need to keep trying.

 I can't do this yet.

See how much you know!

What can I do?			
1 I can understand a bar chart.			
2 I can say what happened in a test.			
3 I can say how objects that make a sound are called 'sources of sound'.			
4 I can name sources of sound.			
5 I can tell someone how a sound gets fainter as it travels away from the source of the sound.			
6 I can use my senses to name sources of sounds.			
7 I can say which part of my body I use to hear.			

I need more help with:

_____

_____

_____

Electricity at home

1 Electricity helps us in our daily lives.

a Draw four things that use electricity.

> Things that use electricity

b Draw four things that do **not** use electricity.

> Things that do not use electricity

Then and now

1 Colour in the boxes in the table to show:
- electrical appliances used today
- objects used before there was electricity.

Two have been done for you.

Appliance	Used today with electricity	Used before there was electricity
vacuum cleaner		
lantern		
dishwasher		
television		
typewriter		
lamp		
whisk		

Attracted to a magnet

1 Which objects will be attracted to the magnet?
Draw a line from the object to the magnet.

leaf

table

steel key

sock

ball

ice cube

plastic button

steel spoon

scarf

iron nail

sweet

steel coat hanger

Do magnets pull (attract) all metals?

A group of learners wanted to find out what happened when they put a magnet near different objects. Some learners thought that all objects made from metal were pulled towards the magnet. Adi and Kim did not agree. They tested objects made only from metal.

Here are their results:

1. Read the sentences below.

   Tick (✓) **yes** if the sentence is correct or cross (✗) **no** if the sentence is not correct.

   a The magnet pulls (attracts) all metals.

   Yes ☐   No ☐

   b The magnet does not pull all metals.

   Yes ☐   No ☐

   c The magnet pulls (attracts) only metals made from iron or steel.

   Yes ☐   No ☐

2. Try out the test at home to find out if magnets pull (attract) all metals.

Down the snakes and up the ladders

1 Play this game with a partner. You need a counter and a spinner. Take turns to spin the spinner and to move your counter.

36	35	34	33	32	31
finish		Oh no! How can we see in the dark with no electricity?			
25	26	27	28	29	30
Oh no! Magnets do not attract plastic.					
24	23	22	21	20	19
	Go up. Magnets attract objects made from some metals.				Go up. Yes, electrical appliances use electricity.
13	14	15	16	17	18
Oh no! Wood is not magnetic.			Go up. A magnet will attract a spoon made from a metal called steel.		
12	11	10	9	8	7
	Oh no! The flashlight does not have a battery, so it does not work.				
1	2	3	4	5	6
start		Go up. Some metals are magnetic.			Go up. Electrical appliances can give out heat and light.

Find a way through a maze

1 You need a magnet and a small magnetic object.

Put the magnet under this page. Put the magnetic object on top of the maze on this page.

Work out the way from START to FINISH with your magnet and object.

Then use a pencil to draw the way through the maze.

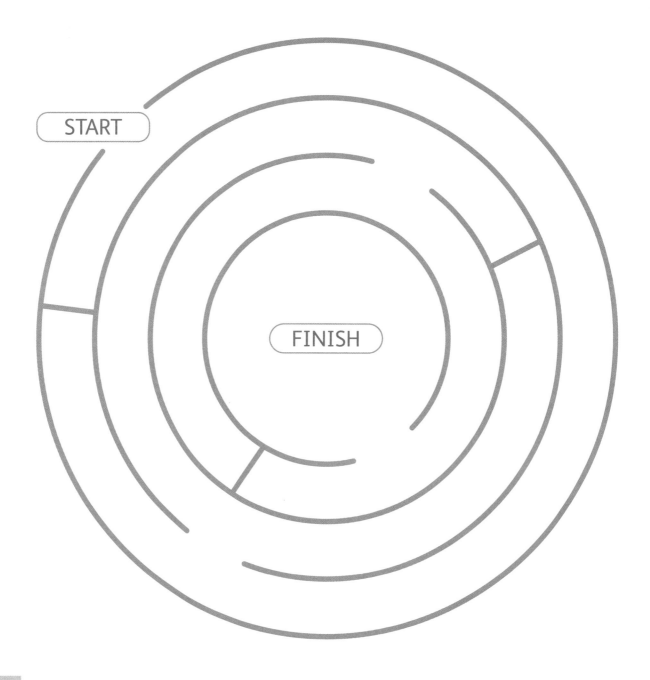

Self-check

See how much you know!

 I can do this.

 I can do this, but I need to keep trying.

 I can't do this yet.

What can I do?			
1 I can sort and group objects.			
2 I can follow instructions when I do practical work.			
3 I can complete a table.			
4 I can tell someone how electrical appliances need electricity to work.			
5 I can name things that need electricity to work.			
6 I can name things that use a battery to work.			
7 I can describe how a magnet attracts objects made from some metals.			
8 I can say how a magnet will not attract objects that are not made from metal.			

I need more help with:

_____

_____

_____

Water everywhere

1 Most of the Earth is covered in water. Match the picture to the correct name. One has been done for you.

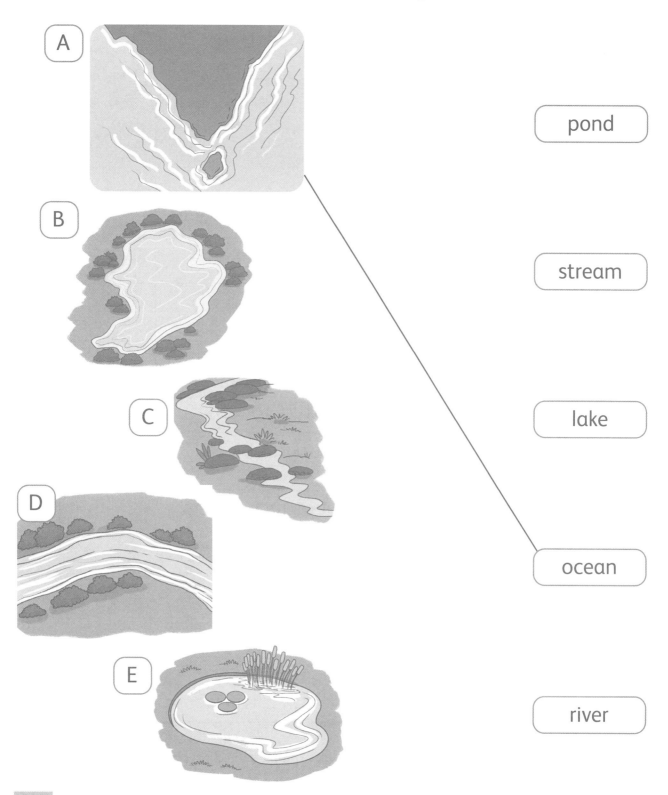

pond

stream

lake

ocean

river

Where do these animals live?

1 Write the correct word next to each animal. Use these words:

river   ocean   pond

A

_____

B

_____

C

_____

2 Complete this sentence. Use one of these words:

chocolate   water   leaves

Most of the Earth is covered in _____.

Do not litter

1 **a** There are objects in the picture that do not belong on the beach. They harm people and animals.

b Cross (✗) the objects that should **not** be on the beach.

Which objects are made from rocks?

1 Tick (✓) the objects that are made from rocks.

diamond

coal

plastic bottle

T-shirt

chair

wall

69

What does a geologist do?

1 Complete this sentence. Use your own words.

A geologist finds out _____

_____ .

2 **a** Draw yourself as a geologist finding out about rocks.

b What are you doing in your drawing?

_____

_____

Objects in the soil

1 Chen dug up soil. He used a magnifying glass to look at the soil.

What do you think he found in the soil?

a In each magnifying glass, draw a picture of what Chen found in the soil.

b Write the names under each picture.

Design a rock garden

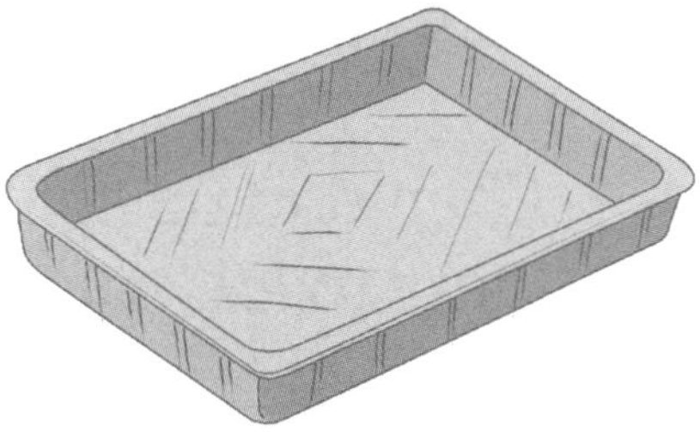

1 **a** What will you use to make your rock garden?

I will need these things:

_____

_____

_____

b What do your plants need to grow?

My plants need these things to grow:

_____

_____

_____

My rock garden design

Unit 8   Planet Earth

Self-check

See how much you know!

 I can do this.

 I can do this, but I need to keep trying.

 I can't do this yet.

What can I do?			
1 I can tell someone that the Earth is mostly covered in water.			
2 I can tell someone about oceans, lakes, rivers, streams and ponds.			
3 I can say what some rocks look like.			
4 I can say what soil is made from.			
5 I can say what different soils look and feel like.			
6 I can say what the land is made from.			
7 I can say what soil is made from.			
8 I can sort rocks into groups.			

I need more help with:

_____

_____

_____

Planets and the Sun

1 Which of these is planet Earth? Circle the correct box.

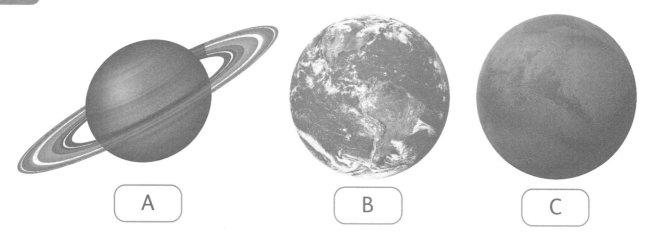

A   B   C

2 Tick (✓) the correct sentences. Cross (✗) the incorrect sentences.

a The Earth is flat. ⬜

b The Earth is a planet in the Solar System. ⬜

c The Earth is mostly covered in water. ⬜

d Humans can live on planet Earth. ⬜

e The Earth gets its heat and light from the Moon. ⬜

3 Tick (✓) the things the Sun gives us.

light ⬜   heat ⬜

sound ⬜   air ⬜

water ⬜

The Sun

1 Jack made a mind map of what he knows about the Sun. Read each sentence. Tick (✓) the sentences that are correct. One has been done for you.

We get light from the Sun. ✓

The Sun is made from ice.

The Sun is a star.

The Sun is a planet.

The Sun is very hot.

The Sun gives us heat.

Look at the stars

1 What is Tony's hobby? Tick (✓) the correct answer.

football   astronomy   reading

2 Draw what Tony looks at through his telescope.

3 Is Tony a scientist? Tick (✓) **yes** or **no**.   Yes ☐   No ☐

4 Draw lines to join up the stars to see the shapes that Tony sees.

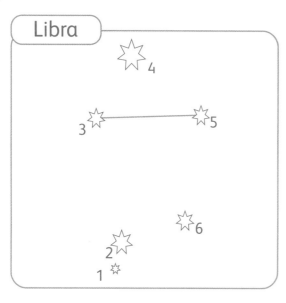

Travel to the Moon

1 The first person to walk on the Moon was Neil Armstrong. He walked on the Moon in 1969.

This is a picture of his space suit.

Label the parts of his space suit. Use these words:

| boots | gloves | oxygen pack | visor |

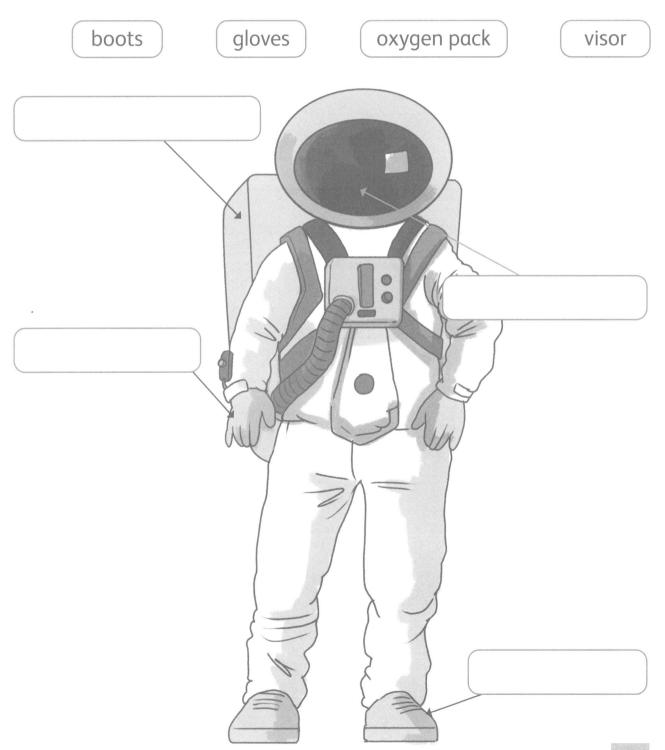

Apollo 11

1 What do you think astronauts took with them when they went to the Moon?

Write seven things:

1.

2.

3.

4.

5.

6.

7.

Design a space rocket

1 **a** Design your own space rocket.

My space rocket

b What will you use to make your space rocket?

Make a list.

_____

_____

_____

_____

_____

_____

_____

_____

_____

_____

2 Ask your partner to write something good about your design.

I like the design because _____

_____.

Name: _____ Date: _____

Unit 9 Earth in space

Self-check

 I can do this.

 I can do this, but I need to keep trying.

 I can't do this yet.

See how much you know!

What can I do?			
1 I can say that we live on the planet Earth.			
2 I can tell someone what the planet Earth looks like.			
3 I can tell someone that the Sun is a star.			
4 I can tell someone that the Sun gives us heat and light.			
5 I can tell someone about the stars in the sky.			
6 I can say what the Earth looks like from space.			

I need more help with:

_____

_____

_____